# River Monsters

Speedy Publishing LLC
40 E. Main St. #1156
Newark, DE 19711

www.speedypublishing.com

Copyright 2014
9781635013320
First Printed November 20, 2014

All Rights reserved. No part of this book may be reproduced or used in any way or form or by any means whether electronic or mechanical, this means that you cannot record or photocopy any material ideas or tips that are provided in this book.

# Alligator gar

The fossil record traces the existence of alligator gars back to the Early Cretaceous over a hundred million years ago. They are the largest in the gar family, and among the largest freshwater fishes in North America.

# Arapaima

The arapaima, pirarucu, or paiche is a genus of bonytongues native to the Amazon and Essequibo basins of South America. They are among the largest freshwater fish in the world, reaching lengths of as much as 4.5 m (15 ft).

# Bull shark

Bull sharks can thrive in both saltwater and freshwater and can travel far up rivers. They have even been known to travel as far up the Mississippi River as Illinois, although there have been few recorded freshwater attacks. They are probably responsible for the majority of near-shore shark attacks, including many attacks attributed to other species.

# The Electric eel

The electric eel (Electrophorus electricus) is a South American electric fish and is known for its strong electric charge discharges measuring around 550 to 600 V, enough to upset other fish perhaps even a human.

# Sawfish

Sawfish can be found both in saltwater or in rivers and creeks, and has been found up to 100 kms inland. Up to 7 meters (23 ft.) in length, sawfish may look like sharks but are actually more closely related to rays. Their "saw" is both a weapon and a sensory organ, covered on electro-sensitive pores which allow it to sense prey despite its terrible eyesight.

# Giant Siamese Carp

Only a fraction of its adult size, this fish is capable of growing to 10 feet and 660 pounds, making it one of the largest species of freshwater fish on the planet.

# Giant Freshwater Stingray

It is found in large rivers and estuaries in Indochina and Borneo, though historically it may have been more widely distributed in South and Southeast Asia. One of the largest freshwater fishes in the world, this species grows upwards of 1.9 m (6.2 ft) across and may reach 600 kg (1,300 lb) in weight.

It has a relatively thin, oval pectoral fin disc that is widest anteriorly, and a sharply pointed snout with a protruding tip. Its tail is thin and whip-like, and lacks fin folds. This species is uniformly grayish brown above and white below; the underside of the pectoral and pelvic fins bear distinctive wide, dark bands on their posterior margins.

# Goonch Catfish

This giant fish can be found in the rivers of south and southeast Asia. The controversy surrounding these fish stem from their interactions with humans along the Kali River, which runs between Nepal and India.

This stretch of river has been used to dispose of funeral pyres after the Hindu funeral rituals. It is believed that the Goonch Catfish has been feeding on these corpses which has allowed them to grow to substantial sizes. As well, it may very well have given them the taste for human flesh.

# Lungfish

Lungfish (also known as salamanderfish) are freshwater fish. Lungfish are best known for retaining characteristics primitive within the Osteichthyes, including the ability to breathe air, and structures primitive within Sarcopterygii, including the presence of lobed fins with a well-developed internal skeleton.

The fins of other lungfishes have become long, wispy sense organs, and they are in general more eellike in appearance. Lungfish feed on snails and plants, storing quantities of fat for sustenance during hibernation.

# New Zealand Longfin eel

The New Zealand longfin eel is a very long lived fish with records of females reaching 106 years old and weighing up to 24 kg. Longfin Eels have the slowest growth rate of any eel species studied, growing between 1 - 2 centimetres a year.

# Vundu Catfish

The vundu is the largest freshwater species in southern Africa, reaching up to 150 cm in length. The maximum recorded weight is 55.0 kg. Few other catfish have such large second dorsal (adipose) fins or such long barbels as do the Vundu. Its barbels nearly reach to the origin of the pelvic fin.

# Piranha

There are approximately 20 species of piranha found living in the Amazon River, with only four or five of them posing any danger. Adult piranha will eat just about anything – other fish, sick and weakened cattle, even parts of people.

# Pacu

Pacu is a South American freshwater fish found in most rivers and streams in the Amazon and Orinoco river basins of lowland Amazonia. Pacu uses its teeth mainly to crush nuts and fruits, but sometimes they also eat other fish and invertebrates. They usually eat floating fruits and nuts that drop from trees in the Amazon, and on a few occasions were reported to attack the testicles of male swimmers mistaking them to be floating nuts.

# The Lake Sturgeon

The lake sturgeon is the largest fish in the Great Lakes and is considered a living fossil because it has survived—virtually unchanged—for more than 150 million years. Lake sturgeon can grow to be more than 300 pounds and live to be more than 100 years old. They can reach a length of 9 feet.

Printed in Poland
by Amazon Fulfillment
Poland Sp. z o.o., Wrocław